SMP 11-16



# Teacher's guide to Book G3

## Contents



CAMBRIDGE
UNIVERSITY PRESS

The School Mathematics Project was founded in 1961 with the purpose of improving the teaching of mathematics in schools by the provision of new course materials. SMP authors are experienced teachers and each new venture is tested by schools in a draft version before publication.

Work on SMP 11–16 started in 1977 and the pilot version of the course has been used by some 50 schools, most of them comprehensive but including some selective schools, since 1980.

Since its inception the SMP has always offered an 'after sales service' for teachers using its materials. If you have any comments on SMP 11–16 or would like advice on its use please write to

SMP office,
University of Southampton
Southampton
SO17 1BJ

The following people have contributed to the writing of the G materials.

Afzal Ahmed
Alex Biddle
Jacqui Bowers
Charles Dickinson
Eric Gower
Harry Gurevitch
Alan Haigh
Graham Hall
Harry Harrison
Bob Hartman
John Hersee
Spencer Instone
Sylvia Johnson
Val Nolan
Jim Prestidge
Geraldine Scanlan
Tony Sharratt
Chris Stolz
Ray Westwood

Others too numerous to be mentioned individually have provided valuable advice and help. Among these are the mathematics staff and pupils of the pilot schools whose detailed comments on the draft version were essential in revising the materials for publication.

The G writing team is led by Spencer Instone.

The SMP 11–16 team is led by John Ling.

# Introduction

SMP 11–16 has been written in the belief that pupils need to be given the time and, through the provision of suitable material, the opportunity to grasp the meaning of mathematical ideas by relating them directly to situations they can understand. Only in this way is it possible for mathematics to become an integral and useful part of their whole experience, rather than a separated activity with its own arbitrary rules, cut off from common sense and the rest of life.

The first part of SMP 11–16 is booklet-based and is described in the Teachers guides to Number, Algebra, Space and Handling data. The second part, for pupils in years 9 to 11 of secondary school, is mainly book-based. This second part provides materials (called Yellow, Blue, Red, Green and Amber) aimed at five levels of ability. The Green, or G, materials have been written for lower ability pupils – those below roughly the 60th percentile of ability. We have taken the view that lower ability pupils need a wide variety in their mathematics – variety of presentation, content, method of working and so on. The G materials reflect this view.

*Books* The G series comprises *Books G1–G9* and *G+*. These are all intended for teacher-led lessons to a group of pupils. *G+* is for use alongside *G2–G7*.

*Booklets* There is a range of topic booklets for use alongside the G books, and also three optional G-booklets for pupils working on *G1, G2* and *G3*. The topic booklets offer students the opportunity to apply their mathematics to real-life situations. They contain questions of a straight-forward nature as well as more open extended option questions. The extended option questions are intended as starting points for pupils to undertake independent problem-solving work on practical topics. (For further details see the teachers notes for each topic booklet.)

*Resource packs* The three resource packs, A, B, and C, comprise games, puzzles and investigations. They are for use by individuals or small groups. Resource Pack A is intended to be used alongside books G1, G2 and G3.

*Worksheets* Most of the books in the series, and some of the booklets, have accompanying worksheets. These are contained in the packs, *Worksheet masters for the G series* (covering Books *G1* to *G7*) and *Worksheet masters for G+, G8* and *G9*.

*Teacher' guides* The teachers guides for *Books G1* to *G9* and *G+* contain not only answers, but suggestions for extending the work in the books, suitable practical activities, areas for further discussion or mental work, as well as pointing out any revision that pupils may need or places where language may be a problem. The guides are a vital part of the G materials.

## Assessment

Appropriate assessment of students can be both informative and motivating. For many students, especially those of lower ability, the prospect of examinations at the end of Year 11 will seem very distant. The SMP Graduated Assessment Profile, going alongside students normal classroom work, has therefore been developed by the SMP and the Oxford and Cambridge Schools Examination Board.

The profile is available for students of all abilities in Years 9, 10 and 11, and comprises assessment items grouped into Stages. The Stages contain items which test written, mental, practical and oral skills. When a student reaches the required standard on (say) Green Stage One, the school can immediately present a Green Stage One Certificate to the student. This bears the Boards name and crests on the front and a summary of the skills covered by the Stage on the rear.

The G materials cover the whole of the National Curriculum in Mathematics up to and including level 5, and the Foundation level of the Scottish Certificate of Education, Standard Grade. Two GCSE syllabi have been prepared specifically based on the SMP 11–16 materials, including the G course.

- Mathematics (SMP 11–16) offers students two terminal examination papers with questions in the same style as the written materials, and includes a coursework component to be completed during Years 10 and 11. For students following the Green course, practical open-ended coursework may be based on the topic booklets.

- Mathematics (SMP Graduated Assessment) offers a modular syllabus. One component of the syllabus is a series of module tests taken during Years 10 and 11. These tests are based on the content of the Stages of the Graduated Assessment Profile. There is a coursework component identical to the Mathematics (SMP 11–16) syllabus, and one terminal paper at the end of the course.

Further details of the Graduated Assessment Profile may be obtained by writing to: SMP Graduated Assessment, Oxford and Cambridge Schools Examination Board, Purbeck House, Purbeck Road, Cambridge CB2 2PU.

Details of the GCSE examinations may be obtained from your MEG Home Office.

# Guide to Book G3

## Equipment and materials for Book G3

*Ancillary materials*

Resource pack A
Booklets: *Pattern*
*Investigating folding*
*Impossible objects*

*Worksheets*

3 worksheets, numbered G3–1 to G3–3
Assessment worksheets (19 sheets)
G3a Recaps 1 and 2
G3b Recaps 1 and 2
G3 Mental test

*Essential equipment*

Ordinary items such as graph paper, squared paper, scissors, card etc. are not included in this list. It is essential that ordinary, four-function calculators are always available.

| | | |
|---|---|---|
| p 12 | Percentage scales | ruler marked in mm and cm |
| p 14 | Flowers | dice |
| p 18 | Area of rectangles | transparent grid of centimetre squares (minimum size about 17 cm by 6 cm, copied onto acetate sheet or perhaps tracing paper. One A4 sheet would then produce about 5 grids) |
| p 33 | Listing | coloured pens or pencils |
| p 35 | Review : angle | angle measurer |

*Ordering equipment and materials*

The following items are published by Cambridge University Press. You should order them through your usual school book supplier.

*Resource pack A* ISBN 0 521 30124 6
*Worksheet masters for the G-series* ISBN 0 521 33624 4
*Topic booklets reference pack*
ISBN
*Impossible objects* (pack of 5 booklets) ISBN 0 521 30306 0
*Investigating folding* (pack of 5 booklets) ISBN 0 521 30305 2
*Pattern* (pack of 5 booklets) ISBN 0 521 30307 9
Angle measurers (pack of 5) ISBN 0 521 25435 3
Pie chart scales (pack of 10) ISBN 0 521 26366 2
Machine stencils (pack of 10) ISBN 0 521 25434 5

When ordering, remember to state the ISBN, the series title (SMP 11–16), the name of the item, the publisher and the number of **packs** you want. (So for example if you want 35 copies of *Pattern*, write your order as '7 packs of 5'.)

Paper circles are required for *Investigating folding*.
Cheap quality filter papers about 110 mm diameter are ideal for this; we suggest you ask your science department about a source of supply.

Items such as dice, counters and mirrors are available from suppliers of educational aids, who issue detailed catalogues.

---

## You are only allowed ...

page 0

*Aims* To encourage simple problem solving abilities, and to give practice in simple number work.

*Special points* You may need to point out that 27 − 3 + 9 is the same as 27 + 9 − 3 etc. The activity lends itself to group work.

*Discussion* The answers to the first investigation are unique: for the second there are many answers. Discussion of different answers will be interesting. Whether or not to allow '33' and '44' in the second part may provoke argument.

You might wish to introduce the idea of brackets to show what needs doing first, in which case many answers shorter than those shown here may be produced.

*Extension* For the first part, you could use 1, 2, 4 ,8 ,16, 32 and just + to form numbers.

For the second part, pupils could investigate whether you can make all numbers under 12 using ×, − and (say) 5 and 3; or (more difficult perhaps) using ÷, + and a suitable pair of numbers.

**Investigation 1**

1 (a) $12 = 9 + 3$ (b) $8 = 9 - 1$
(c) $25 = 27 - 3 + 1$ (d) $17 = 27 - 9 - 1$
(e) $20 = 27 - 9 + 3 - 1$
(f) $14 = 27 - 9 - 3 - 1$
(g) $21 = 27 - 9 + 3$
(h) $34 = 27 + 9 - 3 + 1$

2 All the numbers up to and including 40 can be made.
$1 = 1$ $2 = 3 - 1$ $3 = 3$ $4 = 3 + 1$
$5 = 9 - 3 - 1$ $6 = 9 - 3$
$7 = 9 - 3 + 1$ $8 = 9 - 1$
$9 = 9$ $10 = 9 + 1$ $11 = 9 + 3 - 1$
$12 = 9 + 3$ $13 = 9 + 3 + 1$
$14 = 27 - 9 - 3 - 1$ $15 = 27 - 9 - 3$
$16 = 27 - 9 - 3 + 1$ $17 = 27 - 9 - 1$
$18 = 27 - 9$ $19 = 27 - 9 + 1$
$20 = 27 - 9 + 3 - 1$ $21 = 27 - 9 + 3$
$22 = 27 - 9 + 3 + 1$ $23 = 27 - 3 - 1$
$24 = 27 - 3$ $25 = 27 - 3 + 1$
$26 = 27 - 1$ $27 = 27$ $28 = 27 + 1$
$29 = 27 + 3 - 1$ $30 = 27 + 3$
$31 = 27 + 3 + 1$ $32 = 27 + 9 - 3 - 1$
$33 = 27 + 9 - 3$ $34 = 27 + 9 - 3 + 1$
$35 = 27 + 9 - 1$ $36 = 27 + 9$
$37 = 27 + 9 + 1$ $38 = 27 + 9 + 3 - 1$
$39 = 27 + 9 + 3$ $40 = 27 + 9 + 3 + 1$

3 There is only one way to make each number.

**Investigation 2**

1 There are many different answers. Some suggestions are:
(a) $9 = 3 \times 3$ (b) $12 = 4 \times 3$
(c) $5 = 4 \times 3 - 4 - 3$
(d) $6 = 4 \times 3 - 3 - 3$ (e) $13 = 4 \times 4 - 3$
(f) $17 = 4 \times 3 \times 3 - 4 - 4 - 4 - 4 - 3$ or $33 - 4 - 4 - 4 - 4$
(g) $21 = 4 \times 3 \times 3 - 4 - 4 - 4 - 3$ or $33 - 4 - 4 - 4$
(h) $37 = 4 \times 4 \times 4 - 3 - 3 - 3 - 3 - 3 - 3 - 3 - 3 - 3$ or $44 - 4 - 3$

2 You can make all the numbers under 12.
$1 = 4 - 3$ $2 = 4 \times 3 - 4 - 3 - 3$ $3 = 3$
$4 = 4$ $5 = 3 \times 3 - 4$ or $4 \times 3 - 3 - 4$
$6 = 3 \times 3 - 3$
$7 = 4 \times 4 - 3 - 3 - 3$ or $44 - 33 - 4$
$8 = 4 \times 3 - 4$ $9 = 4 \times 4 - 3 - 4$
$10 = 4 \times 4 - 3 - 3$
$11 = 4 \times 3 \times 3 - 4 - 4 - 4 - 4 - 3 - 3 - 3$ or $33 - 4 - 4 - 4 - 4 - 3 - 3$ or $44 - 33$
$12 = 4 \times 3$

3 Many numbers can be made in more than one way.

# 1 Weight

pages 1–6

*Aims* To give practice in simple manipulations involving weight.
To introduce the equivalences $\frac{1}{4}$ kg = 250 g, $\frac{1}{2}$ kg = 500 g.
Comparing weights using a 'weight line'.
Appreciating that (say) $2\frac{1}{2}$ kg = 2500 g.

*Revision* The simple comparative work in *Book G2* should form a basis for this work. Simple estimation of the weight of a few objects will help place the work in context.

*Language* You will need to draw attention to the number of different alternatives we use when referring to kilograms – kilos and kg. The word 'kilo' is also used in kilometres to mean one thousand, and you could talk about this. In the latter part of the chapter, the word ingredients may need attention.

*Special points* We have not gone as far as '2900 g = 2·9 kg' in this chapter. If you are brave, you might base some work around question E2 in the Home Economics rooms!

*Discussion* Question E1 is very suitable for discussion. The class might be divided into groups and suitable parcels discussed. The list can of course extend beyond the items shown.

*Practical* Different types of scales such as bathroom scales and kitchen scales should be used in conjunction with the chapter, so that pupils get experience with using them. The markings on the scales may well need explaining. Many of the questions might be checked practically. You will have to explain that 1·5 kg is 1 kg 500 g if you weigh 'real' objects, as decimals of a kg is a common way of marking weights over 1 kg. Similarly 'net weight' might need explaining.

*Mental* There is scope for easy mental manipulation of the type: 'A packet of butter weighs $\frac{1}{4}$kg. How many grams do 3 packets weigh?' 'A tin of corned beef weighs 650 g. The empty tin weighs 200 g. How much does the corned beef weigh?' and many more.

*Extensions* Making other weight lines and marking objects on them of interest to the pupils, perhaps with a connecting picture.

---

**A1** (a) 50 kg (b) 57 kg (c) 7 kg

**A2** (a) 128 kg (b) 78 kg

**B1** Less than $\frac{1}{2}$kg

**B2** Less than 1 kg

**B3** 1500 g

**B4** 2500 g

**B5** Less than 1 kg

**B6** Less than 1 kg

**C1**, **C2** (Answers on worksheet G3–1)

**C3** $2\frac{1}{2}$kg is heavier.

**C4** 5 packets of butter

**C5** $1\frac{1}{4}$kg

**C6–C10** (Answers on worksheet G3–1)

**C11** A, Land Rover
B, Double-decker bus
C, Bulldozer
D, Engine

**D1**

| Name | Weight before | Weight after | Weight lost | Money raised |
|---|---|---|---|---|
| Joe | 84 kg | 70 kg | 14 kg | £2·80 |
| Ravi | 90 kg | 66 kg | 24 kg | £4·80 |
| Joan | 78 kg | 48 kg | 30 kg | £6·00 |
| Meg | 66 kg | 52 kg | 14 kg | £2·80 |

**D2** (a) Meg was lightest.
(b) Joan became lightest.
(c) Ravi was heaviest.
(d) Joe became heaviest.
(e) Joan lost most weight.
(f) They raised £16·40 altogether.

**E1** (Old people's parcels)

**E2** (a) 4 eggs (b) 200 g butter (c) 750 g flour

**E3** (a) 48 buns (b) 24 buns (c) 48 buns

**E4** 36 buns (You are restricted by only having 6 eggs.)

---

## Review: enlargement

page 7

**1** (a) $3\frac{1}{2}$ cm
(b) T, $2\frac{1}{2}$ cm  I, $1\frac{1}{2}$ cm  N, 3 cm
(c) T, 7 cm high, 5 cm wide
I, 7 cm high, 3 cm wide
N, 7 cm high, 6 cm wide

**2**, **3** (Enlargements on squared paper)

## 2 Percentage scales

pages 8–13

*Aims* To strengthen understanding of percentages by using scales to read off a percentage of a quantity before doing any calculations.
To give pupils a visual picture of a percentage of a quantity, and thus a method of estimating a percentage.

*Revision* 100% is the whole, 50% half and 25% a quarter of a quantity.
£1·3 means £1·30.

*Language* You may need to discuss or explain the meaning of 'discount'.
In question B3, 'units' (referring to kg, miles, m and so on) may need explanation.

*Discussion* Many questions require an approximate answer. You may want to discuss with pupils how accurate they might expect to get their answers (remembering that printers can never 'register' 2 scales absolutely accurately). The same attention may be given to estimating percentages.

*Estimation/mental* Apart from the estimating work in the chapter itself, you might wish to ask mental questions of the type 'what is 50% of £80' and follow with 'roughly how much do you think 53% of £80 is?'. The diagram on page 13 might be used as a rough check on this.

*Extension* The scale on page 13 might be used to find percentages found in newspapers or mail order catalogues – deposits, discounts, rises etc. The level 3 *Fractions and decimals 2* (pack of 5 booklets ISBN 0 521 27616 0) may be useful.

All the answers in this chapter should be regarded as approximate. Credit should be given for reasonable answers – 'reasonable' will depend on the ability of the pupils and the 'register' of the scales in their text-book.

**A1** £42

**A2** (a) £18 (b) £21 (c) £3

**A3** (a) Just over £19 (b) Just under £35
(c) Just under £44

**A4** (a) Just over 33% (b) Just under 67%
(c) Just over 58% (d) Just under 92%
(e) Just under 57% (f) Just over 28%
(g) 10% (h) Just over 8%

**A5** (a) £6 (b) £66

**A6** (a) £9 discount (b) £51

**A7** A rise of £5 is better, since 7% of £60 is just over £4.

**A8** Dave got the better buy, since 20% of £60 is £12.

**A9** Dawn got the bigger rise since 12% of £60 is just over £7.

**A10** (a) About 16 kg (b) Just under 31 cm
(c) About 5 m (d) Just under 53 tons
(e) Just under 20 gallons (f) 54 litres

**A11** (a) Just under 53 miles
(b) About £5·30

**B1** (Checking that 30% of 150 is 45)

**B2** (a) 90 (b) 15 (c) 120

**B3** (a) Just under 110 kg
(b) Just under 30 miles
(c) Just over 85 m
(d) Between 55 and 60 litres
(e) Just over 90 cm
(f) Between 60p and 65p

**B4** (Checking 26% of 7)

**B5** (a) Just over 3·2 kg (b) 2·1 m
(c) Just over 4·8 km

**B6** (a) £1·40 (b) About £4·40
(c) About 50p

**B7** (a) About £1·10 (b) About £8·10

**B8** £1 off is better since 10% of £7 is 70p.

**B9** It shrinks by about 90 cm or 0·9 m.
It is now about 6·1 m long.

(Section C: You will need to allow **wide** margins of error when marking pupils' answers!)

**C1** Actual answer 22·4

**C2** (a) Actual answer 51·2
(b) Actual answer 9·6 kg
(c) Actual answer £35·84

**C3** (a) £25 (b) £30 (c) 55 kg (d) 55 m

**C4** (a) Actual answer 22·5 km
(b) Actual answer 82·7%
(c) 17·3% – follow through from pupil's answer to part (b)!

**D1** (a) About 3·6 (b) About 4·8
(c) About 4·2 (d) About 10·2

**D2** (a) Pupil's check that 50% of 8 is 4)
(b) 6 (c) About 5 (d) About 0·7

**D3** (a) £4·80 (b) £4·40 (c) About £5·70

**D4** (a) £1·40 (b) £9·10 (c) About £11·80

**D5** (a) About £1·60 (b) About £10·60

**D6** About £2·70

**D7** (a) 48 (b) About 62 (c) About 18

# Flowers

page 14

*Aims* To give some very elementary ideas of simulations.

*Discussion* Of pupils' own rules to maximise flowers will be most useful.

## Review: scales and estimating

page 15

*Special points* The scales are taken from an actual cooker – you may well have to discuss with pupils what each division on a scale stands for before they begin.
Note that the temperature scale is graduated every 25°.

**1** 15 minutes

**2** (a) 13 minutes (b) 18 minutes
(c) 42 minutes (d) 56 minutes

**3** (a) 3:00 (b) 11:15 (c) 1:15 (d) 12:30

**4** A 525°F
B 475°F
C 375°F
D 425°F

## 3 Area of rectangles

pages 16–26

*Aims* The area of a rectangle is found by multiplying its length by its breadth – the chapter shows pupils that this is a measure of the number of centimetre squares inside the rectangle. 'L-shaped' rectangular areas are introduced, and other areas that are easily split into rectangles.

*Language* You will want to emphasise that 'sq cm' is pronounced 'square centimetres' and not 'centimetres squared'. Unless the distinction is made, it is unclear whether (say) '12 centimetres square' refers to a shape 12 cm by 12 cm, or one for example 2 cm by 6 cm.

*Special points* The chapter requires a transparent grid of centimetre squares. The minimum size for this is about 17 cm by 6 cm, though it is better if larger. A master might be copied onto acetate sheet or perhaps tracing paper. It would probably be unwise to ask pupils to make one of these, as the grid must be accurate.

The abbreviation $cm^2$ for square centimetres is introduced later in the course.

*Discussion* In section D, you may wish to discuss the fact that different ways of splitting the L-shape do not produce different areas.

*Practical* It may be possible to measure the classroom to the nearest metre and find its area (the area of the ceiling may be less cluttered). Different shaped rooms may be compared. Rough sketches may help if it is necessary to split up the shape.

Areas of other objects may be found – a page of book G3 (measure in cm), the blackboard (measure in feet?), windows, the desk surface etc.

Questions might be asked like 'This 1 litre tin of paint covers 15 square metres. How many tins would we need to give the ceiling a single coat?' You might then consider carpeting the calssroom!

*Estimation/ mental* Direct estimation of area is very difficult, and not to be recommended. However, rough estimates of the 2 lengths involved, and then multiplying in the head is good practice.

*Extension* Level 3 *Area 2* (pack of 5 booklets ISBN 0 521 27602 0) covers much the same ground, but also includes area of a triangle.

---

**A1** (a) 6 sq cm (b) 12 sq cm (c) 8 sq cm (d) 9 sq cm (e) 12 sq cm (f) 4 sq cm

**A2** (a) 15 sq cm (b) 20 sq cm (c) 24 sq cm (d) 16 sq cm (e) 72 sq cm

**B1** (a) 32 sq cm (b) 42 sq cm (c) 15 sq cm (d) 18 sq cm

**B2** E has the largest area.

**B3** (a) 6 sq cm (b) 9 sq cm (c) 15 sq cm (d) 6 sq cm (e) 24 sq cm (f) 16 sq cm (g) 12 sq cm (h) 18 sq cm (i) 4 sq cm

**C1** (a) 4 rows (b) 5 squares (c) 20 sq cm

**C2** (a) 3 rows (b) 7 squares (c) 21 sq cm

**C3** 16 sq cm

**C4** (a) 12 sq cm (b) 16 sq cm (c) 12 sq cm (d) 28 sq cm

**C5** (a) 6 by 3 (b) 18 sq cm

**C6** 27 sq cm

**C7** A, 30 sq cm E, 40 sq cm
B, 32 sq cm D, 45 sq cm
C, 36 sq cm

**D1** (a) 8 sq cm (b) 4 sq cm (c) 12 sq cm

**D2** (a) 3 sq cm (b) 12 sq cm (c) 15 sq cm

**D3** (a) 16 sq cm (b) 14 sq cm

**D4** 15 sq cm

**D5** (a) $6 + 6 = 12$ sq cm
(b) $10 + 2 = 12$ sq cm

**D6** $9 + 10 = 19$ sq cm
or $16 + 3 = 19$ sq cm

**E1** (a) (Pupil's plan)
(b) (c) (The areas of pupil's rectangles)
(d) 14 square metres

**E2** (a), (b), (c) (Pupil's plan, split up)
(d) 20 square metres (e) £180

**E3** (a), (b), (c) (Pupil's plan, split up)
(d) 26 square metres

**E4** (Splitting in different ways)

---

## Review: lines, points and distances

page 27

*Revision* You may need to revise this method of drawing parallel lines with the pupils before they attempt this exercise.

---

**1** 3 sets

**2** 7 mm

**3** (Pupil's flag)

---

# Review: negative numbers

page 28

**1** (a) $^{-}3 + 4 =$ (1) (b) $^{-}6 + 5 =$ ($^{-}1$)
(c) $^{-}2 + 5 =$ (3) (d) $^{-}8 + 2 =$ ($^{-}6$)
(e) $^{-}1 + 7 =$ (6) (f) $^{-}9 + 9 =$ (0)
(g) $^{-}10 + 11 =$ (1) (h) $^{-}4 + 9 =$ (5)
(i) $^{-}5 + 10 =$ (5)

**2** (a) $^{-}5 +$ (8) $= 3$ (b) $^{-}7 +$ (3) $= {}^{-}4$
(c) $^{-}8 +$ (12) $= 4$ (d) $^{-}3 +$ (3) $= 0$
(e) $^{-}1 +$ (11) $= 10$ (f) $^{-}2 +$ (1) $= {}^{-}1$
(g) $^{-}6 +$ (9) $= 3$ (h) $^{-}4 +$ (4) $= 0$
(i) $^{-}9 +$ (18) $= 9$

**3** (a) $3 - 0 =$ (3) (b) $^{-}2 - 4 =$ ($^{-}6$)
(c) $5 - 10 =$ ($^{-}5$) (d) $^{-}1 - 3 =$ ($^{-}4$)
(e) $^{-}2 - 6 =$ ($^{-}8$) (f) $^{-}7 - 4 =$ ($^{-}11$)

**4** (a) $5 -$ (11) $= {}^{-}6$ (b) $4 -$ (7) $= {}^{-}3$
(c) $3 -$ (13) $= {}^{-}10$ (d) $6 -$ (9) $= {}^{-}3$
(e) $10 -$ (14) $= {}^{-}4$ (f) $7 -$ (9) $= {}^{-}2$

**5** $^{-}3$°C

**6** $^{-}7$°C

# 4 Listing

pages 29–33

*Aims* When pupils of this ability try to make exhaustive lists of possibilities, they often approach the problem haphazardly. They do not try to be systematic. This is fine as long as there are only a few cases to be considered, but it does not work well when there are a larger number. The object of this chapter is simply to introduce the idea of systematic listing. There is no theoretical or abstract approach. The object is simply to list possibilities so that you can be sure you have not missed any.

*Special points* The use of a letter notation for easy recording should be encouraged. Questions C2 and C3 present different tabulation problems.

*Discussion* The first part of section A (p. 29) lends itself well to group or class discussion. The question 'How can we be sure we have all the possible ways?' can be used to introduce a reason for systematic working.

Discussion of how to shorten the task may lead to the use of single letters. You may well want to discuss the answers to most of the questions in the chapter.

*Practical* The use of coloured counters to represent soup, pudding and fish and chips may help with the introduction.

**A1** (Answers in text on page 30)

**A2** (Checking there are no other ways)

**A3** (a) Cereal, eggs or toast.

(b)

| | | |
|---|---|---|
| C | E | T |
| C | T | E |
| E | C | T |
| E | T | C |
| T | C | E |
| T | E | C |

(C = cereal,
E = eggs,
T = toast)

**A4**

| *1st* | *2nd* | *3rd* |
|---|---|---|
| E | S | W |
| E | W | S |
| S | E | W |
| S | W | E |
| W | S | E |
| W | E | S |

**A5** (a)

| *1st* | *2nd* | *3rd* |
|---|---|---|
| A | B | C |
| A | C | B |
| B | A | C |
| B | C | A |
| C | A | B |
| C | B | A |

(b) There are 6 different ways.

**A6** There are 24 ways.

**A7**

| | | | |
|---|---|---|---|
| m | s | i | c |
| m | s | c | i |
| m | i | s | c |
| m | i | c | s |
| m | c | i | s |
| m | c | s | i |

**A8** (a)

| | | | |
|---|---|---|---|
| i | m | s | c |
| i | m | c | s |
| i | s | m | c |
| i | s | c | m |
| i | c | s | m |
| i | c | m | s |

(b)

| | | | |
|---|---|---|---|
| c | m | s | i |
| c | m | i | s |
| c | s | m | i |
| c | s | i | m |
| c | i | m | s |
| c | i | s | m |

**A9** 24 ways

**B1**

| *1st* | *2nd* |
|---|---|
| A | B |
| A | C |
| A | D |

**B2** (a)

| *1st* | *2nd* |
|---|---|
| B | A |
| B | C |
| B | D |

(b)

| *1st* | *2nd* |
|---|---|
| C | A |
| C | B |
| C | D |

(c)

| *1st* | *2nd* |
|---|---|
| D | A |
| D | B |
| D | C |

(d) 12 ways

**B3**

| *1st* | *2nd* |
|---|---|
| A | B |
| A | C |
| A | D |
| A | E |
| B | A |
| B | C |
| B | D |
| B | E |
| C | A |
| C | B |
| C | D |
| C | E |
| D | A |
| D | B |
| D | C |
| D | E |
| E | A |
| E | B |
| E | C |
| E | D |

20 ways altogether

# Review: decimals

page 34

**1** 0·14 cm

**2** 0·29 cm

**3** 0·43 cm

**4** (a) 0·06 (b) 0·25 (c) 0·41
(d) 0·59 (e) 0·8 or 0·80 (f) 0·96

**5** (a) 4·16 (b) 2·99 (c) 1·75
(d) 1·59 (e) 1·87 (f) 10·02
(g) 9·98 (h) 4·5 or 4·50
(i) 0·30 or 0·3 (j) 0·75

## Review: angle

page 35

| | | |
|---|---|---|
| **1** A, DC8<br>B, Tristar<br>C, Concorde<br>D, TU144<br>E, Boeing 747 | **2** 12 and 3<br>1 and 4<br>2 and 5<br>3 and 6<br>4 and 7<br>5 and 8<br>6 and 9<br>7 and 10<br>8 and 11<br>9 and 12<br>10 and 1<br>11 and 2 | **3** *a* and *h*<br>*b* and *d*<br>*e* and *g* |

## 5 Carrying on

pages 36–42

*Aims* To extend the use of 2 decimal places to include measuring in m and cm. To give practice in using a calculator with m and cm in the contexts of addition and subtraction.

*Revision* It would probably be useful to look at 2 decimal places on a number line as a reminder before beginning the chapter.

You may need to ask how many cm there are in a metre.

*Language* The phrase 'as a decimal of a metre' on page 39 may need explaining.

The pronunciation of 'hundredths' (as opposed to 'hundreds') may again need emphasis. On page 40 we use the term 'units' referring to the whole number nearest the decimal point – make sure there is no confusion between this and its use to mean 'metres', 'kilometres' etc. The word 'surveyor' may need attention.

*Discussion* You may want to discuss why 2 m 5 cm is 2·05 m, and not 2·5 m – one suggestion is given on page 38. Another might be to think of a digital measurer, whose digits measure metres, tenths of metres and hundredths of metres.

On page 41, the equivalent statements

$$100\,\text{cm} = 1\,\text{m}$$
$$1\,\text{cm} = 0{\cdot}01\,\text{m}$$

may need talking over with the pupils.

In question C12(c), the halfway point between 1·21 and 1·22 is mentioned. This is a 'throw-away' introduction to the third decimal place, which you might want to follow up.

*Practical* Using real surveyor's tapes may be possible (note that the end of the tape is likely to be different from the illustration at the top of page 36). In this case measuring a length in cm and then using the tape will help reinforce the 'decimal of a metre – centimetre' conversion.

*Mental work* Pupils should be able to tackle conversions between m and cm in their heads by the end of the chapter.

---

**A1** (a) 95 cm (b) 98 cm

**A2** (a) 0·63 m long (b) 0·66 m long

**A3** (a) 1·37 m (b) 5·23 m (c) 3·96 m

**A4** (a) 4 cm longer (b) 104 cm
(c) 1·04 m = (104) cm

**A5** 109 cm

**A6** (a) 102 cm (b) 205 cm (c) 309 cm
(d) 110 cm (c) 120 cm

**A7** 250 cm

**A8** (a) 2 m 60 cm (b) 1 m 80 cm (c) 70 cm
(d) 2 m 62 cm (e) 1 m 85 cm (f) 75 cm

**A9** (a) 1·9 or 1·90 m (b) 2·8 or 2·80 m
(c) 2·75 m (d) 2·85 m
(e) 2·1 or 2·10 m (f) 2·05 m

**A10** (a) 1·06 m (b) 1·01 m (c) 2·04 m
(d) 2·09 m (e) 3·08 m (f) 1·03 m
(g) 0·03 m

**A11** (a) The downstairs ceiling by 34 cm
(b) 54 cm (c) 25 cm

**B1** (a) 3·54 m or 3 m 54 cm
(b) 2·81 m or 2 m 81 cm
(c) 4·09 m or 4 m 9 cm
(d) 1·87 m or 1 m 87 cm

**B2** (a) 0·44 m or 44 cm
(b) 1·42 m or 1 m 42 cm
(c) 1·65 m or 1 m 65 cm
(d) 1·89 m or 1 m 89 cm
(e) 0·82 m or 82 cm

**C1** 1·08

**C2** (a) 1·05 (b) 1·01

**C3** 1·5

**C4** (a) 2·3 (b) 2·03
(c) 4·06 (d) 4·6

**C5** 1 m 2 cm = 1 and (2) hundredths of a metre, or (1·02) m

**C6** 1 m 4 cm = 1 and (4) hundredths of a metre, or (1·04) m

**C7** (a) 1 m 7 cm (b) 1·07 m

**C8** 1·09 m

**C9** (a) 1·2 or 1·20 m (b) 2·1 or 2·10 m
(c) 3·2 or 3·20 m

**D1** (a) 2·1 m is bigger (b) By 2 cm

**D2** (a) £3·60 (b) £18

**D3** (a) 2 m 9 cm (b) 1 m 82 cm
(c) 3 m 80 cm (d) 2 m 40 cm

**D4** (a) 40 cm (b) Sharon, by 35 cm

**D5** (a) 1·5 m (b) 1·35 m
(c) 1·215 m

---

## 6 Stacking and filling

pages 43–48

*Aims* To provide an informal introduction to the ideas behind the volume of a rectangular block.

To give further practice in the multiplication of 3 or more numbers together.

*Special points* Volume of a block as such is not introduced.

*Language* 'Pallet' will need explaining.

*Discussion/ practical* The connection between multiplication and repeated addition can be made when looking at a stack of (say) four layers. To some pupils, it may seem unreasonable at first sight that such large answers are obtained when finding the number of boxes in a stack. Doing a few examples using wooden blocks or centicubes would be useful.

In section D, as an introduction, pupils might try working out how many packets of chalk (say) will fit in a given cardboard box. The packet itself can be used to see how many will fit in the length, width and height of the box. After this, measurement might be used to see how many packets fit in the 3 directions, rather than moving the packet.

You may well get a different number of packets into the box if you change its orientation.

*Estimation* Estimation of volumes is notoriously difficult; estimating the number of boxes or packets in a pile can probably only be done by the inexperienced by multiplying. It may still be valuable to try to estimate, however, even if only to show how bad we are at these sorts of estimates!

*Extensions* For many pupils section D itself may be an extension! You might also want to look at the number of boxes packed in 'cubes' of sides 1, 2, 3, 4 and so on, and note how quickly the number of boxes in the cubes grows.

Level 3 *Squares and cubes* (pack of 5 booklets ISBN 0 521 27606 3) covers much of this work, and also introduces some notation.

---

**A1** (a) 9 packets (b) 16 packets

**A2** (a) 9 (b) 12 (c) 16

**A3** (a) 20 (b) 30 (c) 20

**B1** 27 boxes

**B2** (a) 16 boxes (b) 48 boxes

**B3** (a) 12 boxes (b) 2 layers (c) 24 boxes

**B4** (a) 60 (b) 45 (c) 60 (d) 32 (e) 12

**B5** (a) 30 (b) 36 (c) 20 (d) 24

**B6** 180 boxes

**B7** 100 boxes

**C1** 60 boxes

**C2** (a) 32 boxes (b) 128 boxes

**C3** (a) 45 boxes (b) 24 boxes (c) 135 boxes (d) 72 boxes

**D1** 4 packets along the box

**D2** 4 packets across the box

**D3** 96 packets in the box

**D4** (a) 10 layers (b) 6 packets (c) 6 packets (d) 360 packets

**D5** 60 packets

**D6** (a) 90 (b) 64 (c) 90 (d) 84

**D7** (a) 336 boxes
(b) He will get the same number in this way round.

---

## Magic squares

page 49

*Aims* To encourage practice of simple number skills in an easy 'problem solving' situation.

*Discussion/extension* The solutions to questions 1, 2 and 3 and to questions 4, 5 and 6 are essentially identical. You might want to bring this out in discussion. In the same way adding, say, 1 to each number, or multiplying each number by 10 in a magic square will produce a square that is still magic. You might find it interesting to discuss what other magic squares might be found in this way.

---

**1**

| | | |
|---|---|---|
| **6** | 7 | 2 |
| **1** | 5 | **9** |
| 8 | **3** | 4 |

**2**

| | | |
|---|---|---|
| **8** | **3** | **4** |
| 1 | 5 | 9 |
| **6** | **7** | 2 |

**3**

| | | |
|---|---|---|
| 6 | **1** | **8** |
| 7 | 5 | **3** |
| **2** | 9 | 4 |

**4**

| | | |
|---|---|---|
| 67 | **1** | **43** |
| 13 | 37 | **61** |
| **31** | **73** | 7 |

**5**

| | | |
|---|---|---|
| 67 | **13** | **31** |
| **1** | **37** | **73** |
| 43 | 61 | 7 |

**6**

| | | |
|---|---|---|
| 43 | **61** | **7** |
| 1 | **37** | 73 |
| 67 | **13** | **31** |

---

## 7 Have you the time?

pages 50–55

*Aims* To introduce the terms a.m. and p.m. and the 24-hour clock.

*Language* You may want to point out that in the 24-hour system, 1500 is pronounced 'fifteen hundred' or 'fifteen oh oh', but never one thousand five hundred or 15 o'clock.

*Revision* The work in book G1 on the 'time line', and the equivalence of, for example, 2:45, 15 minutes to 3, and a quarter to 3 may need revision.

*Discussion* Questions C4, C5 and C6 are suitable for class discussion – who got most or least sleep, watched most or least television etc. You might discuss how the lengths of time vary at weekends or in the holidays.

It may be useful to discuss why anyone uses the 24-hour clock at all – where it is most useful (it will be taken up in timetabling work later in the course). It is notable that some transport companies have reverted to 'a.m.' and 'p.m.' as causing less confusion to most people!

*Practical* If there is a source of timing devices, pupils can be asked to set them to (say) 3 o'clock in the afternoon. Central heating clocks, automatic timers for electrical sockets, video tape recorders and other devices often use the 24-hour clock.

*Mental work* Simple intervals can be worked mentally. ('It is a quarter to three now; the train goes at 1527 – how long is that?').

There are many questions which ask the pupil to write down an equivalent time in 24-hour form or using a.m. or p.m. which are suitable for asking mentally.

---

**A1** (a) 8:00 a.m. (b) 10:00 a.m.
(c) 10:00 p.m. (d) 10:30 p.m.
(e) 2:00 a.m. (f) 3:00 p.m.

**A2** You might have lunch (or dinner) at 12:45 p.m.

**A3**

| | |
|---|---|
| 7:30 a.m. | Got up |
| 8:00 a.m. | Breakfast |
| 8:25 a.m. | Left for school |
| 8:45 a.m. | Arrived at school |
| 10:45 a.m. | Break started |
| 12:20 p.m. | Dinner |
| 3:45 p.m. | School ended |
| 4:05 p.m. | Got home |
| 5:00 p.m. | Tea |
| 7:30 p.m. | Started homework |
| 8:45 p.m. | Watched T.V. |
| 10:30 p.m. | Bed |

**A4**

| | |
|---|---|
| 9:05 a.m. | 2:05 p.m. |
| 10:05 a.m. | 3:05 p.m. |
| 11:05 a.m. | 4:05 p.m. |
| 12:05 p.m. | 5:05 p.m. |

**A5** (a) 5:20 p.m. (b) 5:50 a.m.
(c) 9:15 p.m. (d) 3:45 a.m.

**B1** 57 minutes

**B2** 8:30 p.m.

**B3** (a) 1010 (b) 2310 (c) 1630 (d) 0735

**B4** (a) 3:20 p.m. (b) 4:35 p.m.
(c) 7:03 a.m. (d) 10:07 p.m.
(e) 6:47 p.m. (f) 11:24 a.m.
(g) 1:09 p.m. (h) 11:59 p.m.

**B5** (a) 1610 (b) 2015 (c) 0631 (d) 2142
(e) 2217 (f) 2318 (g) 0936 (h) 1937

**B6** (a) Steve must use the footbridge.
(b) Rachel cannot buy a ticket.
(c) It is closed.
(d) It is open for $3\frac{1}{2}$ hours.
(e) She can collect the parcel.
(f) The office is open.

**C1**, **C2** } (Answers on worksheet G3–2)

**C3** (a) 1 hour (b) 45 min
(c) 1 hour 45 min

**C4** (a) 15 min (b) 3 hours 45 min
(c) 45 min (d) 8 hours 15 min

**C5**, **C6**, **C7** } (Pupil's own day on worksheet G3–3)

---

## Review: fractions

page 56

1 (b)

2 (a) $\frac{3}{4}$ (b) $\frac{1}{2}$ (c) 4 times

3 (a) 250g (b) 750g

4 (a) 30km (b) Jim (c) 15km (d) £2·25

## Review: chapters 1 and 2

page 57

1 6 bags

2 1300g

3 950g

4 (a) 250g (b) 125g (c) 500g

5 (a) 50% (b) 25%

6 (a) 3 tonnes of gravel
(b) 3 tonnes of cement, 6 tonnes of sand

7 (a) About 900g (b) About 675g
(c) 10 to five, or 4:50

## Review: chapter 3

page 58

1 (a) 9sq cm (b) 8sq cm (c) 16sq cm
(d) 15sq cm (e) 6sq cm (f) 10sq cm

2 (a) 18sq cm (b) 13sq cm

3 (a) 32sq cm (b) £384

## Review: chapters 4 and 5

page 59

1 There are 20 ways for the 2 prizes to be won.

| | | | | | |
|---|---|---|---|---|---|
| 1st | AAAA | HHHH | CCCC | PPPP | BBBB |
| 2nd | HCPB | ACPB | AHPB | AHCB | AHCP |

2 There are 15 ways if Red–Blue is counted the same as Blue–Red and Green–Green etc. are allowed.

3 3m 55cm (b) 96cm

4 (a) £5 (b) £5·50 (c) £7
(d) £3·00
(e) They measure 6m 5cm. The charge is £7.

5 (a) This bus will go under.
(b) A $4\frac{1}{2}$m bus will not go under.

## Review: chapters 6 and 7

page 60

1 (a) 18 (b) 12 (c) 36 (d) 30
(e) 30 (f) 18

2 (a) 1900 (b) 1800 (c) 1830 (d) 2015
(e) 2055 (f) 2225 (g) 2354
(h) Flash back

# Guide to assessment for Book G3

## Recaps G3a and G3b

**Suggestions for use**

1 There should (within reason) be no time limit for a pupil working a Recap.
2 You may read any question to a pupil who has reading difficulties.
3 70% would represent a reasonable 'pass' mark. (Allow one mark for each part of each question. For marking purposes, no account should be taken of units, including £ or p, unless otherwise stated in the answers. You may wish to comment to pupils if they miss out or use incorrect units.)
4 If a pupil gets less than 70% you should alow at least a fortnight for review and revision before a pupil retakes the parallel Recap.

## **G3a Recap 1** (covers G3 pages 1–28)

**1** (a) 45 kg (b) 63 kg
(c) 18 kg (Follow through from (a) and (b).)
(d) 4 kg (Follow through from (a))

**2** (a) More (than $\frac{1}{2}$ kg)
(b) 700 grams (c) More (than 1 kg)

**3** 1 mark for each arrow which you consider the pupil intends to point to the correct mark on the scale.

**4** (a) Accept answers between 4·1 and 4·2 inclusive.
(b) Accept answers between 2·4 and 2·5 inclusive.
(c) Accept answers between 7·05 and 7·15 inclusive.
(d) Accept answers between 3·3 and 3·4 inclusive.

**5** (a) Accept answers between 1·55 and 1·65 inclusive.
(b) Accept answers between 7·35 and 7·45 inclusive. (Follow through from (a).)

**6** (a) 36 miles
(b) Accept 70% marked with limits shown

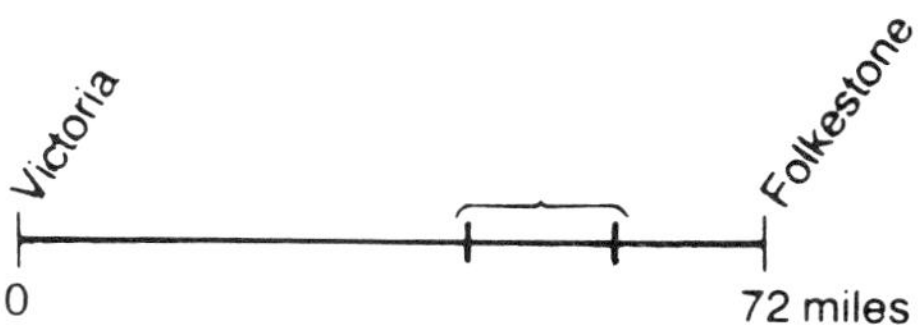


(c) Accept 40 to 65 (Follow through from (b).)

**7** (a) 24 sq m (b) 26 sq m (c) 32 sq m
(d) £96 (Follow through from (c).)

**8** (a) 4 sets (b) (2 times enlargement)
(c) (Accurate drawing of flag)

## G3a Recap 2 (covers G3 pages 1–28)

**1** (a) 320 g (b) 440 g
(c) 120 g (Follow through from (a) and (b).)
(d) 260 g (Follow through from (a).)

**2** (a) Less (than 1/4 kg)
(b) 500 g (c) 680 g

**3** 1 mark for each arrow which you consider the pupil intends to point to the correct mark on the scale.

**4** (a) Accept answers between 56 and 57 inclusive.
(b) Accept answers between 18 and 19 inclusive.
(c) Accept answers between 37 and 38 inclusive.
(d) Accept answers between 22 and 23 inclusive.

**5** (a) Accept answers between £11·50 and £12·50 inclusive.
(b) Accept answers between £65·50 and £68·50 inclusive. (Follow through)

**6** (a) £24
(b) Accept 65% marked within limits shown

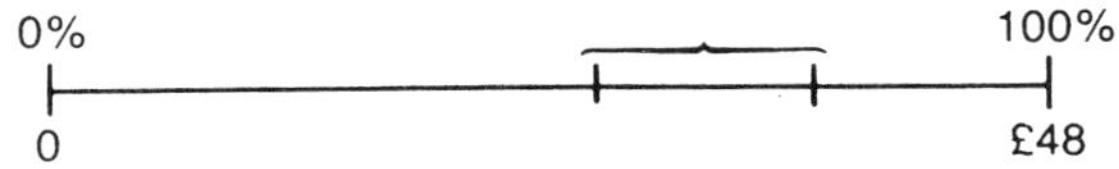


(c) Accept 26 to 38 (Follow through from (b).)

**7** (a) 30 sq cm (b) 36 sq cm
(c) 41 sq cm (d) 72 sq cm

**8** (a) a is parallel to b
c is not parallel to d
e is parallel to f
g is not parallel to h
(b) (Pupil's flag)

## G3b Recap 1 (covers pages 29–end)

**1** (a) 1·12 m (b) 1·99 m (c) 1·06 m

**2** (a) 1·06 m or 1 m 6 cm or 106 cm
(b) 2·13 m or 2 m 13 cm or 213 cm
(c) 1·92 m or 1 m 92 m or 192 cm

**3** (a) 7·16 m (b) 7·46 m
(c) 7·6 or 7·60 m (d) 7·81 m

**4** (a) 10 minutes
(b) 4 hours
(c) Yes
(d) Morning
(e) 10 15 or 10:15
(Accept any suitable separator between 10 and 15.)

**5** (a) 36 (b) 24 (c) 45 (d) 20

**6** (a) 36 (b) 96

**7** 1 mark for making a list that is in general systematic.
2 marks for listing 20–24 different sandwiches, including the 2 given.
1 mark for listing 16–19, including the 2 given.

## G3b Recap 2 (covers pages 29–end)

**1** (a) 3·11 m (b) 3·99 m (c) 4·89 m

**2** (a) 1·78 m or 1 m 78 cm or 178 cm
(b) 1·37 m or 1 m 37 cm or 137 cm
(c) 2·02 m or 2 m 2 cm or 202 cm

**3** (a) 4·15 m (b) 4·39 m
(c) 4·7 or 4·70 m (d) 5·02 m

**4** (a) 3 hours
(b) 10:30 p.m. or 10 30 p.m.
(c) Evening
(d) 19 05 or 19:05
(Accept any suitable separator between 19 and 05.)
(e) $3\frac{1}{2}$ (hours)

5 (a) 30 (b) 28 (c) 60 (d) 30

6 (a) 105 (b) 60

7 1 mark for a generally systematic list.
2 marks for listing 15–20 different placings including the 2 given.
1 mark for listing 10–14 different placings including the 2 given.

---

# G3 Mental test

**Suggestion for use**

1 Prepare and duplicate an answer sheet for each pupil.
2 On a copy of the mental test choose and underline the alternative you will read out for each question.
3 Read each question slowly and distinctly twice. Any numbers should be read out correctly in full. (For example, read 126 as 'one hundred and twenty six', not 'one two six'.)
4 Allow enough time for pupils to answer (as a guide, between 5 and 10 seconds.)
5 Allow 1 mark per question (take no account of units, including £ and pence); 70% would represent a reasonable 'pass' mark.
6 If a pupil gets less than 70%, you should allow at least a fortnight for review and revision before he or she retakes the mental test.

# Suggested practical and oral test - Weight: notes and answers

**Essential equipment**

A set of kitchen scales which the pupil is familiar with.
A pack of Kellogg's Variety cereals; select 3 packets with differing weights.
4 objects, each weighing less than 1 kg, not marked with their weights.
About 500 g of rice, dried beans, drawing pins or anything else suitable for question D below.

**Suggestions for use**

1 You will need to test pupils individually. All your questions are given orally, and the pupil gives his or her answers orally.
2 We suggest that the pupil needs to pass all 4 sections to pass the test. Criteria for passing are given below.

**Instructions**

**A** Indicate the three Varieties packets to the pupil. Ask him or her 'Which of these packets says it weighs seventeen grams?' (use one of the three differing weights)

**B** Give the pupil one of the other Varieties packets. Ask 'How much does it say this packet weighs?'

**C** Indicate the 4 objects. Tell the pupil that you want the objects put in a row in order of weight, lightest first. Point out that the pupil can use the scales to help.

**D** Indicate the rice (or equivalent) to the pupil. Ask them to use the scales to weigh out exactly 100 grams (or 150 grams, or 200 grams or 250 grams) of it.

**Marking**

**A** Pass if the pupil indicates the correct packet.
**B** Pass if the pupil tells you the correct weight, including using the word 'grams'. Do not prompt for 'grams' if the pupil omits it or says 'seventeen gee' or other incorrect verbalisation.
**C** Pass if the pupil puts the objects in the correct order.
**D** Pass if the pupil weighs the rice or equivalent to a reasonable degree of accuracy.

Published by the Press Syndicate of the University of Cambridge
The Pitt Building, Trumpington Street, Cambridge CB2 1RP
40 West 20th Street, New York, NY 10011–4211, USA
10 Stamford Road, Oakleigh, Melbourne 3166, Australia





First published 1985
Sixth printing 1996

Printed in Great Britain by Watkiss Studios Ltd, Biggleswade, Beds.

ISBN 0 521 31660 X

ED